Tesla Cybertruck

Revolutionizing the Automotive Industry

(How Elon Musk and Electric Cars Changed the World)

Bobby Edwards

Published By **Elena Holly**

Bobby Edwards

Tesla Cybertruck: Revolutionizing the Automotive Industry (How Elon Musk and Electric Cars Changed the World)

ISBN 978-1-9994523-1-5

No part of this guidebook shall be reproduced in any form without permission in writing from the publisher except in the case of brief quotations embodied in critical articles or reviews.

Legal & Disclaimer

The information contained in this book is not designed to replace or take the place of any form of medicine or professional medical advice. The information in this book has been provided for educational & entertainment purposes only.

The information contained in this book has been compiled from sources deemed reliable, and it is accurate to the best of the Author's knowledge; however, the Author cannot guarantee its accuracy and validity and cannot be held liable for any errors or omissions. Changes are periodically made to this book. You must consult your doctor or get professional medical advice before using any of the suggested remedies, techniques, or information in this book.

Upon using the information contained in this book, you agree to hold harmless the Author from and against any damages, costs, and expenses, including any legal fees potentially resulting from the application of any of the information provided by this guide. This disclaimer applies to any damages or injury caused by the use and application, whether directly or indirectly, of any advice or information presented, whether for breach of contract, tort, negligence, personal injury, criminal intent, or under any other cause of action.

You agree to accept all risks of using the information presented inside this book. You need to consult a professional medical practitioner in order to ensure you are both able and healthy enough to participate in this program.

Table Of Contents

Chapter 1: The Story of Its Founders

Martin Eberhard's and Marc Tarpenning's stories are inextricably linked to Tesla Motors. It was their love of sustainable transportation combined with an entrepreneurial spirit that laid the groundwork for one the most innovative, groundbreaking companies in history.

Martin Eberhard is a former software engineer who loves electric cars. In particular, he was intrigued by the possibility of using electric vehicles to reduce dependency on fossils fuels while combating climate change. Marc Tarpenning is an electrical engineering who shares Eberhard's love of sustainability. Marc was also eager to offer his technical knowledge.

Eberhard Tarpenning teamed up with Eberhard in 2003 and launched Tesla Motors. It was a mission that seemed simple, yet ambitious. Their goal: To accelerate the global transition to renewable energy.

Eberhard Tarpenning pursued their vision despite many difficulties in the beginning, especially the challenge of designing a completely new car. Their complementary skills and faith were used to help them overcome hurdles, keep Tesla Motors on track, and continue their journey.

Tesla Motors continues Martin Eberhard's and Marc Tarpenning's legacy, as their unwavering passion for sustainability and pioneering spirit continue to push the company to new heights. Tesla today is one of the leaders in the electrified vehicle market, and it's a force behind the transition to more sustainable living. This story of Tesla's founders is testimony to the power and potential of innovation, entrepreneurship, as well as the will to transform the world.

Elon Musk joins Tesla

Tesla Motors started out as a simple idea. It quickly evolved into one the most innovative, exciting and successful companies in the auto industry. Elon Mots was one of those who played an important role in the success.

Elon was appointed chairman of Tesla's board in 2004 and became its product architect. In 2008, he then took over as CEO. Tesla at the time was a new startup which had released its very first product: the Tesla Roadster. The company benefited from his unique experience, energy, and vision.

Musk not only served as the chairman of his company, but he also put in money and was actively involved with product development. He collaborated closely with engineers to design the Model S, an electric luxury sedan of high performance that is the best available on the marketplace.

Tesla under Elon has been a leader in innovation, pushing what is thought to be possible for the automobile industry. Elon's vision was to not only revolutionize the automobile industry, but accelerate the global transition to renewable energy.

Martin Eberhard (co-founder of Tesla Motors) and Marc Tarpenning (co-founder of Tesla Motors) stepped aside from their respective roles in 2008 as a result differences between the CEO Elon and his management style. The co-founders of Tesla Motors, Martin Eberhard and Marc Tarpenning, stepped down from their roles in 2008 due to differences between the CEO Elon Musk's management style and strategy.

Eberhard and Tarpenning's departure marked a pivotal moment in Tesla's history. The Roadster was one of their most important contributions.

Elon has had a huge impact on the success of Tesla. Elon's involvement with Tesla has been a key factor in its success. He was crucial to securing financing, directing product development, and driving overall company strategy. Also, he is a big advocate for electric vehicle. He works hard to promote the importance of environmentally friendly transportation.

The entrepreneurial spirit he brings, along with his visionary thoughts and relentless drive for success, has been instrumental in his accomplishments. Tesla under his direction has been a leader for sustainable transportation, and is helping shape the future automotive industry.

Elon Musk - The visionary at the wheel

Elon Musk immediately comes to your mind when thinking about Tesla Motors. South African-born Elon has contributed to the success and expansion of Tesla, as well as the mission the company is on.

Elon Musk is among the most creative entrepreneurs in our time. Elon has transformed multiple industries by creating solutions that are sustainable and have the capacity to revolutionize the world. Elon began his fascination with technology, entrepreneurship and the South African culture at a very young age. In 1972, Elon started programming computers when he was 12 years old. He sold his very first videogame at 24.

Elon Musk's first selling game was Blastar, a game with an outer space theme. At the age of 12, he created the video game and sold it at a price around $500. It was published by a PC and Office Technology Magazine. The BASIC program was used to create a simple game in which he had the player defend their ship.

Elon has pursued interests in internet, finance and renewable energies since moving to the United States. Zip2, the company that supplied city guides to newspaper, was founded by Elon. It was later acquired by Compaq for approximately $300 million. He later founded X.com a company for online payments that was eventually acquired by PayPal. eBay bought it in 2006 for $1.5billion.

Musk is best known as a co-founder and the creator of PayPal, which has become a widely used online payment service. But his ambitions went beyond the tech world. Musk founded Tesla Motors (in 2003) with a passion to promote sustainable energy.

Musk was the driving force behind Tesla's effort to produce stylish electric vehicles which were powerful and exciting to drive. The Tesla Roadster and Model 3 are just two of the vehicles that Musk played an important role in creating.

Musk's role is not limited to Tesla. He also co-founded SpaceX. The company was created to facilitate space exploration. Musk has also been a supporter of renewable energy. His role was crucial in the creation of Tesla's Energy Storage Systems, which were designed to distribute and store energy.

Musk, in spite of the obstacles and challenges Tesla has experienced over the last few decades, has never wavered from his dedication to the mission of the company.

Elon is an incredible visionary. His impact on the entire world can't be emphasized enough. Tesla Motors is his vehicle to transform the automobile industry. It has inspired an entire generation of young entrepreneurs to be ambitious and to make positive impacts on the planet.

Early Challenges

Tesla's early days: Navigating its challenges

The task of starting a brand new company can be challenging, and if you are aiming to revolutionize the automotive industry by doing so it will take even more effort. Tesla Motors' mission in 2003 was one that faced many obstacles and challenges. Tesla founders were determined to make their dream of electric cars a reality, even though they had many challenges to face. Here, we will look closely at

Tesla's first challenges and their solutions.

Electric Vehicles: Lack of Knowledge and Skepticism

Tesla had to deal with a lot of skepticism and lack of knowledge about electric cars in the beginning. Electric vehicles were viewed as unreliable, slow and inefficient at that time. Tesla had a hard time convincing people to invest in their electric cars. Tesla however was determined that people would change their perceptions about electric cars and they were just as good as gasoline-powered vehicles.

Tesla made a big effort to educate the public about the many benefits electric vehicles offer. The company held events to test electric vehicles and gave out information about them. Also, they talked about their advantages with everyone who would listen. Tesla managed to slowly, but surely overcome skepticism in order to create a community who are passionate about electric cars.

Charging Infrastructure Not Available in All Areas

Tesla also faced a major obstacle in its initial years: the scarcity of charging infrastructure. In order for electric vehicles to function, they require a large network of chargers. However, at the time such a system was almost nonexistent. Electric vehicles were unable to be driven long distances, and this lowered their appeal. Tesla wanted to solve this problem, too, so it developed its Supercharger network.

Superchargers make long-distance driving in electric vehicles easy. Charging stations strategically placed near major highways allow drivers to easily and quickly recharge their cars when taking a short break. Supercharger was a large undertaking for Tesla, but essential for its success. The network is today one of largest charging networks worldwide.

Established Automakers - Competition

Tesla was up against established automakers. Many of them were already making investments in electric vehicles. This was a group of companies with deep pockets and an abundance of resources that were determined not to lose their grip on the automobile industry. Tesla had to work hard to stay competitive, but it was able overcome the challenge through innovation. The company built vehicles that were quicker, more efficient and fun to drive.

Tesla didn't just copy other automakers. Tesla always pushed the envelope and tried new things. Its innovative spirit is what sets it apart. Its electric vehicles are not only environmentally-friendly and efficient, but also fun to operate. They appeal to new generations of car purchasers.

Manufacturing and Production challenges

Tesla's early years saw significant production and manufacturing challenges. Tesla was creating a completely new car and learning along the way. It was a big challenge to find a way to mass produce electric vehicles. Tesla overcame this challenge through persistence and hardwork. It is now one of the biggest and most successful electric car manufacturers on the planet.

Tesla has overcome many of the challenges it faced in the early days and set the stage for its success. The hard work of the company founders paid off when they created sustainable energy and innovative electric vehicles. Tesla's initial years were difficult, but essential for its growth.

Tesla Roadster launches: Disrupting status quo

Electric Vehicle Market: The Impact of the Electric Vehicle Market

Tesla Roadster, which was introduced for the first time in 2008 was an absolute game changer. Tesla Roadster is the first Tesla produced car. It was Tesla's bold declaration of its mission to change the status quo in the automobile industry. This is exactly what it did! Roadster had an impressive design and performance. With cutting-edge technologies, it was unlike other electric vehicles.

Electric Vehicles: A Challenge to Perception

The Roadster changed all that. Before, most people thought electric vehicles to be slow, unreliable, and costly. Roadster changed everything. This fast and stylish car proved that an electric vehicle could be as good or better than a gasoline-powered automobile. This car challenged a lot of preconceived notions that people had about electric cars. They showed them to be just as good, if not better than gasoline-powered vehicles.

Electric Cars Made Accessible

Tesla Roadster's greatest contribution was to make electric cars available to more people. Roadster's high-performance sports cars were priced competitively with similar vehicles of its type, proving that electric cars are a viable and cost-effective option. With its long range, fast charging and virtually maintenance free driving, the Roadster was a hit with car buyers.

Tesla as an Industry Leader

Roadster is not just great car for Tesla. It's also an important milestone. Tesla became a market leader for electric vehicles with the launch of its Roadster. It also set up the company to be successful in the future. Tesla Roadster proved to the public that it was a serious manufacturer of high-quality, electric cars.

Innovating the Future

Roadster represented more than just an automobile; it offered a glimpse of the transportation future. Roadster proved that electric cars are just as practical, fashionable, and enjoyable as regular cars. Roadster's impressive range and fast charging abilities set the standards for what is possible on the electric vehicle marketplace.

Inspiring a Movement

Tesla Roadster has had an impact that is far-reaching. This launch inspired car buyers of a younger generation to look at electric vehicles and helped set the foundation for the future growth of this market. Roadster proved that electric cars were just as convenient, reliable, and enjoyable, as other cars.

Roadster's Legacy

Tesla Roadster has remained a symbol, even today, of Tesla's goal to revolutionize automobiles and introduce electric vehicles into the mainstream. His impact was huge on the electric car market, and other electric vehicles followed suit. Roadster is a car that continues to impress and excite buyers across the globe with its performance and innovative technologies.

Tesla Motors has been a pioneer in electric vehicle technology since the launch of its Roadster. Roadster was able to challenge many preconceived ideas about electric cars and make them accessible for a larger audience. Tesla became a leading player in the market for electric cars with the launch of its Roadster, which set the tone for their continued success. Roadster's legacy inspires car shoppers to this day.

Innovation in Action

Chapter 2: The Development and Release

Model S (Model X), Model Y (Model Y) and Model 3

Tesla's innovation is at the heart of their business. Their goal was to reinvent the auto industry through electric vehicles which not only met, but also exceeded, the performance and style of gas-powered cars. Model S is the epitome of that goal. The Model X, Model Y, and Model 3 have all had a positive impact on automotive technology in their respective ways.

Model S Elevating Electric Vehicles To New Heights

Model S was Tesla's first foray into the high-end luxury car segment, and quickly made a name for itself. Model S was a game-changer for Tesla, as its interior design, technology and performance were all of a high standard. Model S' quiet and smooth ride was loved by drivers. Its innovative features including the 17" touch screen panel control, Autopilot capability, and Tesla Supercharging made it popular for daily commutes and long distance trips.

Tesla Model S, first launched in 2012, has become one of today's most technologically-advanced vehicles. There are many features in the Tesla Model S that improve driving performance and enhance safety.

Model S features an electric engine that delivers instant torque, smooth acceleration, and a quiet ride. Model S has a wide range of options for batteries, such as a 75-kWh battery, which can go up to 371, and even 402 miles, on a single battery charge.

Model S features a large touchscreen display of 17-inches that controls most car features including climate control and audio system. Navigation is also included. This intuitive interface makes it easy for drivers of all skill levels to get the information and controls that they need.

Model S comes packed with safety features. It has received five stars in safety from National Highway Traffic Safety Administration. Autopilot helps the driver to maintain control over the vehicle, helping him avoid an accident. There is also a complete airbag system, providing maximum safety in case of a crash.

Model S features a variety of advanced technologies, which are intended to enhance energy efficiency and minimize the environmental impact. Model S is equipped with a sophisticated regenerative braking technology that uses the energy usually lost in braking to recharge its battery. This car also has an aerodynamic body that reduces wind resistance to increase range.

Tesla Model S, as a whole, is an advanced car that has many features designed to make driving more enjoyable and sustainable. Model S can be a great option for anyone who wants a cutting-edge electric car. It's also a good choice for those looking to buy a safer and greener vehicle.

Model X: A Fusion between Style and Function

Model X was Tesla's next major project. This car pushed back the limits of the automobile industry. Model X's distinctive design was topped with advanced technology and unique features like falcon wing doors. Model X was also a popular choice among families because of its impressive interior space and high-tech features. Model X is a vehicle that offers a smooth, quiet drive thanks to its all-electric motor. The advanced safety functions, such as the low center of gravitation and Autopilot features make this one of the safer vehicles available on the road.

Model X electric luxury SUV has a roomy and open interior, thanks to iconic Falcon Wing door. Model X boasts a long range. Some models are able to go up to 370 miles in a single charging. Model X has impressive acceleration. Models are able to accelerate from 0-60 mph 2.8 sec. Model X includes a host of safety technologies, such as an advanced airbag, automatic emergency braking technology and collision prevention.

Model Y : compact SUV with great inside

Model Y is Tesla's latest SUV, an electric compact SUV. Model Y uses the same Platform as Model 3; it has a spacious cabin and is capable of all-wheel driving. Model Y models are also capable of going from 0-60 mph (in 3.5 seconds) and boast impressive range. Tesla Autopilot's latest technology is included, and allows you to drive hands free in certain situations.

Model 3 Electric Vehicles: Affordability and Accessibility

Tesla's Model 3 represented its greatest challenge: to design an electric vehicle accessible to the largest audience. Model 3's sleek design, innovative technology and attractive price made this goal a reality. Model 3 quickly became popular and one of most sought-after vehicles. Model 3 has been designed for the everyday driver. It's easy to drive and is practical. Model 3 is a good choice for drivers who are looking to cut their carbon emissions without having to sacrifice convenience.

Model 3 features a sleek, modern design and offers impressive acceleration and range. Model 3s can reach up to 358 mph and go 0-60 in 3.2

secs. Model 3 vehicles are also fitted with Tesla Autopilot which enables hands-free operating in certain situations.

Tesla Model Y: the Newest Addition to Tesla's Lineup. Potential Impact of the Automotive Industry

Tesla Model Y was unveiled in September of last year. The public has eagerly awaited it since then. This flexible SUV was designed to provide the convenience of an automobile with the performance and efficiency that Tesla is known for. Model Y is a revolutionary vehicle with its large interior space, stylish design, and high-performance range.

Model Y offers a wide range. This electric car has the best range on the current market with up to a 326-mile range. Tesla's battery-technology expertise and its dedication to maximizing energy consumption are responsible for this impressive range. Model Y has Tesla's signature performance acceleration. It can accelerate from 0-60 mph in just 3.5 seconds.

Model Y's impressive performance is matched by its practicality. Seating up to seven people and offering ample storage, the Model Y SUV is perfect for families who want more space than a typical SUV. This SUV also has a panoramic roof which lets in natural light. It creates an airy and spacious feel.

Model Y also focuses on sustainability. As with all Teslas, the Model Y produces no emissions and is 100 percent electric. It means drivers don't have to be concerned about the impact they are having on the environment. They can ride in comfort and peace without any worries. Tesla's Supercharger network allows drivers to quickly recharge their Model Y while on long drives, ensuring there is no need for a stop at the gas station.

Model Y has also been equipped with Tesla's advanced Autopilot System, which allows for semi autonomous driving. This system utilizes cameras and sensor to offer driver assistance and, in some circumstances, can completely take over driving. Model Y comes with Tesla's most recent infotainment systems, featuring a large central touch screen and voice activated controls.

Model Y represents a welcome addition to Tesla's range and is poised to be a game changer in the auto industry. This combination of sustainability, performance and practicality will make it a popular choice for electric car enthusiasts. Model Y can be a good family vehicle, or a performance-oriented electric SUV.

Innovation with a Legacy

Tesla's Model S, Model X, Model 3, and Model Y all demonstrate its dedication to innovation and sustainable development. Tesla's Model S, X and Model 3 have set new standards for electric vehicles. And they have encouraged a younger generation of buyers to look at electric vehicles. Tesla's vehicles are known for their performance, style, and innovative technology. They have become the leader of the automotive industry, as well as the symbol of the transportation of tomorrow.

Tesla's lineup of vehicles is designed to change the way people think about transportation. Each model offers its own set of features. But they all have one goal in common: to revolutionize this industry. Tesla's line of vehicles, with their impressive acceleration and range to the cutting-edge self-driving features and advanced safety technologies is meant to transform how people view transportation. Tesla is able to provide a car that meets your requirements and goes beyond your expectations.

Tesla's success story would not be complete without its Model S, Model X,

Model 3, and Model Y. The vehicles are a testament to Tesla's dedication to innovation and sustainable development, as well as a challenge for many people who have preconceived ideas about electric vehicles. Tesla has become a symbol for the future of transport with its stylish, high-performance vehicles. Tesla's successes have shown that they can produce electric cars that are as good, or even better, than the traditional ones. They also prove that there is a future for electric transportation.

Tesla's Quest for Lower Production Prices

Elon Musk started Tesla Motors (now Tesla) in 2003 with a vision. Not only did he want to revolutionize automotive technology, but he also wanted the costs to be brought down so that they would become more accessible for all customers. It has become a focus area for Tesla Motors over the last few years. They have also made huge strides.

Tesla reduces production costs through the use of advanced technology in its factories. In order to improve efficiency and decrease waste, the production process at Tesla is heavily dependent on robots. Not only has this improved efficiency but also the cost of producing electric vehicles.

Tesla reduced their production costs in part by designing and manufacturing its own parts. The company has reduced production costs by taking charge of its supply chain. By eliminating middlemen and controlling quality, it was able to cut down on the price. The ability to quickly introduce new functions and abilities to vehicles has enabled them to be more competitive than those who have to rely on third-party vendors.

Tesla's efforts to cut the costs of its batteries are also part of these initiatives. Electric vehicles are dependent on this component, which is why lowering its cost can have a huge impact. Tesla has reduced the cost of their batteries by a significant amount over the years.

Tesla's production has been streamlining to decrease waste and boost efficiency. In order to improve efficiency and decrease waste, the company was able to find and remove bottlenecks.

Chapter 3: Tesla Giga Press

Tesla Giga Press a large production tool that is used to produce electric vehicle (EV), batteries. IDRA Group of Italy developed the cutting-edge tech. They are recognized as a leader in high-tech automotive production systems.

IDRA Group has been a part of Piovan Group for over 40 years. Piovan Group is the global leader of production of rubber and plastic auxiliary equipment. IDRA Group is a company with over 40 years' experience. It has built a reputation in providing its clients with innovative and high-quality products.

IDRA Group has proven its expertise with the Tesla GigaPress. The machine was specifically designed to manufacture large battery cells which are a vital component of EVs. The press has advanced control and automation systems that allow it to produce high-quality batteries with consistency and efficiency.

IDRA Group has a commitment to both sustainability and environmental responsibility. The company has a commitment to use renewable energy resources and reduce its carbon foot print in all manufacturing processes. Tesla Giga Press reflects the focus of this company on sustainability.

IDRA Group's partnership has proven to be extremely successful. Giga Press plays a crucial role in production of Tesla Model 3 - a highly acclaimed vehicle. The IDRA Group's partnership with Tesla is extremely successful. The Giga Press played an important role in the production of Tesla Model 3 which has received high praise.

Tesla Giga Press' cutting-edge technology represents the future for automotive manufacturing. The Tesla Giga press is a huge hydraulic presses that produces some of Tesla's critical parts, like the housings for front and rear electric engines, the drive unit, and battery casings.

The Giga Press is a symbol of Tesla's innovation and commitment to sustainability. It also represents its firm belief in a bright future. Tesla was able to use cutting-edge tech to drastically reduce the energy and materials needed to make these crucial components.

One of the biggest features of the Giga Press, is its sheer size. A massive machine, it stands at over six metres tall and is 400 tonnes heavy. Its size is not a problem, as it can still produce parts with tolerances of less than 0.50 millimeters.

Giga Press also impresses with its high speed. It can make a casting complete in just a few moments, which is one of the fastest manufacturing processes in the industry. Tesla is able to lower the production costs by reducing time and increasing efficiency.

Tesla pushes the limits in automotive manufacturing with its Giga Press. The Giga Press, which uses cutting-edge technology and artificial intelligence to create vehicles more energy efficient as well as more sustainable, is helping Tesla. Tesla's leadership in automotive manufacturing will continue to be the norm for years to come.

Tesla's efforts to lower production costs have been key in the company's success. In order to meet the needs of the customers, Tesla relies on advanced technology to control its supply chain.

Tesla Production Sites

Powering Future Automotive and Energy Industries

Tesla, the leader in electric cars and clean energy, has created several production centers around the world. This is to keep up with the demand of its products. Tesla has equipped these sites with the latest technology and innovative methods to guarantee that each vehicle and energy solution produced by Tesla are of high-quality.

The key Tesla sites are described in this chapter. You will learn more about their production, locations, and products.

We'll start by taking a quick look at Tesla's Fremont factory. Model S's, Model X's, Model 3's, and Model Y's are made here. Fremont's factory was initially established by NUMMI. This joint venture between General Motors (GM) and Toyota. Tesla acquired the location in 2010 and transformed it since into a facility that produces high-tech, electric vehicles. Over 10,000 people are employed at the site, which has a floor area totaling 5.3million square feet. Fremont's manufacturing process is highly automated. Advanced technology and robots are used for each vehicle.

Tesla, along with its Californian production center, also has a facility in Shanghai. Built to meet the demand for electric cars in the Asian markets, the factory started producing Model 3s in 2019. Shanghai's factory is spread over an expanse of more than 1.4million sq. feet. It has an annual capacity to produce 250,000 cars. Tesla's Shanghai facility is the country's first foreign-owned factory. The factory is also a sign of Tesla's determination to serve customers worldwide.

Tesla's Gigafactory located in Nevada is also an important Tesla manufacturing site. This facility is solely dedicated to manufacturing battery cells. It is also one of largest buildings. It has features that make it highly sustainable and efficient, such as solar panels on the roof. The Gigafactory manufactures the lithium-ion batterie that powers Tesla's electric vehicles, energy storage systems and more. With a production capability of over 35 GWh a year.

Tesla's production facilities are vital to its business success. They also play a key role in its mission to revolutionize automotive and energy sectors. Tesla's production facilities feature cutting-edge tech, innovative process, and highly-trained employees. These are testaments of Tesla's dedication to creating high-quality products for customers across the globe. Tesla is a global leader in the production of electric vehicles, as well as clean energy.

Tesla's gigafactories at Berlin and Texas

Tesla has always been on the cutting edge of technological innovation. It is pushing the limits of what can be done with energy storage systems, electric vehicles and renewable energy. Tesla's desire for innovation led to the establishment of a number gigafactories throughout the world. These include two large facilities located in Berlin (Germany) and Austin (Texas).

Tesla Gigafactory, located in Berlin in Germany, is Tesla's company's European first factory. The facility will play a vital role in Tesla efforts to increase the speed of the switch to sustainable energy. The Tesla factory in Berlin, Germany will produce electric vehicle, battery, and energy-storage systems. It is a major hub for Tesla's European operation. Tesla's Berlin Gigafactory, which will be located in Berlin, Germany's capital city of Berlin's suburbs and districts in Berlin-Brandenburg and Berlin-Adlershofen in Berlin's district Friedrichshain/Berlin, is expected to bring with it the latest manufacturing technology and the production techniques that made Tesla vehicles and products the most effective and efficient on the market.

Austin, Texas will host the Tesla Gigafactory. It is the primary manufacturing facility of the Model Y Crossover SUV. Texas will be a crucial part of Tesla's ecosystem for energy, producing batteries and storage systems. Texas' gigafactory will focus on sustainability and energy efficiency. It is also equipped with wind and solar turbines.

Tesla is investing heavily in its future, for both sustainable energy sources and vehicles. These gigafactories show that the company has a commitment to leading innovation and bringing about positive change. Tesla will be playing a crucial role with the Berlin, Texas, and other gigafactories in its efforts to bring the world a cleaner and more sustainable future.

Tesla's Revolutionary Batteries: The Development of Tesla's Revolutionary Batteries

Innovative Impact: A journey of innovation

Battery is at the core of an electric vehicle's performance. The story of Tesla's battery development will be explored in this chapter. These batteries have changed the standard of electric vehicles.

Tesla Motors' early days were marked by the focus on battery packs and electric powertrains. Co-founders Martin Eberhard & Marc Tarpenning put their entire effort in creating a powerful, long-range sports car. Tesla Roadsters of the 2008 model year were equipped with battery packs that provided 245 mile range on a charge. At the time, the average EV range was about 100 miles.

Tesla did not want to remain a mere leader in the industry of electric cars. In order to reach a wider market, the company needed to improve battery performance while reducing costs. Tesla Energy, a division that focuses on developing and producing stationary energy storage systems as well as commercial-scale battery packs was born.

Tesla Model S made a major breakthrough when it was introduced. It was fitted with 18650 batteries, which were smaller, lighter, more dense, and energy efficient than the Roadster battery packs. Model S has a 300-mile range, which makes it the largest electric vehicle in the world. Tesla's new battery was more cost-effective and efficient, allowing it to sell a high-performance vehicle for a reasonable price.

Tesla has continuously innovated and improved its battery technologies over the years. The company developed a proprietary system for managing battery packs that is safe and effective. They have also invested heavily in battery research, development and production, such as the Giga Factory. The Giga Factory's capacity is greater than that of the lithium-ion cells produced annually worldwide in 2013.

Tesla's latest technology in battery cells is the Model 3 2170 cell. The largest ever battery cell and with its impressive energy density it allows for an improved range. Tesla uses this cell in stationary products like the Powerwall or Powerpack. This makes it a versatile, scalable and adaptable technology.

Tesla's history of battery development demonstrates the commitment that it has to innovation, sustainability and a greener world. Tesla's early electric car development paved the way for an environmentally friendly future. Tesla redefined electric vehicle standards with its battery-powered technology. This has made EVs a more attractive choice for consumers and set up a new era in sustainable transportation.

Tesla 4680 cell - latest battery technology

Tesla has been working hard to innovate the energy storage industry and automobiles. The Tesla 4680 Battery Cell is one of their most advanced and revolutionary developments. Tesla's Gigafactory is home to a brand new, in-house designed cell that has the ability to improve Tesla's electric cars' performance.

This cell's dimensions are 46mm dia. and 80mm. Tesla previously used 18650 and 217 cells. This new design, however, offers many advantages over those older cells. First, because the 4680 has a bigger form factor than older cells, more energy can be stored. This leads to EVs with a larger range and higher energy density. They also weigh less and cost less.

A 4680 cell's improved thermal control is an important feature. A larger cell allows for better heat dissipation. This reduces any risk of thermal problems and runaway. This results in a battery with a longer life and higher efficiency.

Tesla already began incorporating 4680 cells in its products. This cell is already used in Tesla's new Model S & Model X models, with more to come. Tesla is working with these cells to develop energy storage solutions, like residential or commercial battery systems.

Overall, Tesla's 4680 battery represents an important step forward in developing energy storage systems and electric vehicles. Tesla is aiming to improve the energy density of its 4680 cell, increase range and enhance thermal management.

Driving the FutureTesla's dedication to self-driving technologies

And it's Impact on Transportation

Tesla has always led the way in terms of innovation and technology. The company's dedication to developing self-driving vehicles is one of their most notable achievements. Tesla's dedication to creating self-driving vehicles dates back to the beginning of their company.

Autopilot: the Beginning

Tesla released its Autopilot system in 2015. Since then, it has become one of

Tesla's most-talked about and popular features. Autopilot combines GPS, cameras, ultrasonic and radar sensors with GPS in order to guide the vehicle on highways, roads, etc. without any driver interaction. The technology was updated multiple times after its initial introduction. Now, it includes features such as Autopark and Summon.

Future of Transportation: Self-driving Cars

Tesla's goal is the creation of fully self-driving cars, and it has made significant progress. Tesla is investing heavily to develop its self driving technology. Each software update makes Tesla's cars more capable. Tesla's self-driving car technology will increase road safety, reduce congestion, and enhance the overall efficiency in transportation.

Chapter 4: Self-Driving Technology Influences

Tesla's focus on self-driving technologies is having a major impact in the automotive industry. Tesla's success is inspiring other manufacturers to invest heavily in their self-driving tech. Tesla's achievements in this field are inspiring an entire new generation of technologists and engineers to take up careers as autonomous vehicle specialists.

Tesla's auto-driving tech has had a profound impact not just on the industry of automobiles, but on our society as well. When vehicles can drive by themselves, it will give us more time for other tasks and allow us to travel faster and safer. It will be easier for elderly or disabled people to move around the city and take part in society with self-driving cars.

Tesla's innovative self-driving vehicle technology is having an impact in both the automotive and the society. In addition, it inspires a young generation of engineers to consider careers as autonomous vehicles. Tesla leads the field in self-driving cars and the future transport with its advanced technology.

Tesla Artificial Intelligence

The Road to an Intelligenter Future

Tesla Motors' forays into AI are no exception. Tesla's autopilot and AI-powered production processes are just two of the ways it is harnessing AI power to revolutionize automotive manufacturing.

Tesla's AI-powered autopilot system is perhaps the best known example. The system uses ultrasonic sensors with radar and cameras to guide vehicles on roads without human assistance. Thanks to Tesla's extensive fleet of vehicles, the autopilot system continuously learns and gets better. Along with improving the driving experiences of Tesla owners, autopilot also helps to build the foundations for fully autonomous vehicle.

Tesla isn't the only automaker that has embraced AI. AI can also be used to optimise the manufacturing process, which reduces production times and increases efficiency. Tesla, for instance, employs AI algorithms to analyse the production process and find bottlenecks. It can make changes in real time. As a result, production times have been reduced and the manufacturing process has become more efficient.

Tesla's AI use doesn't end there. AI technology is being used to improve the efficiency and effectiveness of energy storage by Tesla. Tesla's continued expansion in the renewable energy industry is a major factor. This will allow it to continue to help businesses and households with clean and sustainable energy.

Tesla's embrace AI shows the commitment of the company to drive innovation and shape the future of automotive industry. Tesla's embrace of AI is just one example of its commitment to driving innovation and shaping the future of the automotive industry.

Tesla's AI-powered Journey to Self Driving Cars

Tesla's mission is to provide sustainable energy to the world and improve its quality. This goal goes beyond producing electric vehicles. Tesla has been making significant progress in the development of autonomous driving technology. The goal is fully self-driving cars. Tesla has used artificial intelligence as an important component in their effort. We will examine how the company uses AI to meet this goal.

Tesla Autopilot is a new technology.

Tesla Autopilot offers advanced driver assistance features, including adaptive cruise control, lanekeeping, and automated parking. The system collects information using ultrasonic sensors and radar and AI algorithms analyze it to arrive at decisions. Autopilot, which uses AI to make the driving experience safer and better, can improve the overall driving experience.

Computer Vision and Safe Driving

Tesla Autopilot relies on an important component of computer vision. The camera on the car captures images of the road, surrounding area, and the computer vision algorithm processes these images in order to detect and identify objects such as traffic signals, road signs, or other vehicles. AI systems use this data to help them decide on how to move the vehicle.

It can, for instance detect an orange traffic light to automatically slow or stop your vehicle. Also, it can identify other vehicles in the area and utilize this information for maintaining a distance. Tesla Autopilot uses computer vision to provide safety and convenience not possible with traditional driver assistance.

Deep Learning-The Power of Experience

Deep Learning is another AI technology critical to Tesla Autopilot. Deep learning algorithms help the car learn from a vast amount of information, like driving patterns and behavior, to continually improve its performance. As the vehicle drives, its knowledge and ability to adapt to new situations improves.

In order to learn, the vehicle will need to be able to respond and predict to different scenarios. For instance, it can anticipate how to merge onto an interstate or navigate through congested traffic. The algorithms for deep learning can be continuously updated, adding new information and learning to the system. This will ensure the vehicle is always at the forefront in autonomous driving.

Full Autonomy, the Future of Driving

Tesla's ultimate aim is full autonomous vehicles where the driver can just sit back, relax and the vehicle will do all the driving. Tesla is combining AI technologies including deep learning, computer vision and control system to accomplish this goal.

Computer vision algorithms give the vehicle a 360° view of its environment. This allows it to identify and avoid obstacles. Deep learning algorithms constantly improve the ability of vehicles to make decisions, by learning from their experiences and adapting them to different scenarios. Control systems are responsible for managing the vehicle's interaction and movement with the surroundings, optimizing the performance and ensuring safety.

Tesla's AI use has placed the company in front of the self driving industry. It has also allowed the company to continually push the boundaries for autonomous vehicles. Tesla's desire to achieve a more sustainable future is evident in the company's decision to use AI technology to make driving easier and safer.

Tesla's AI technology in the development of self-driving vehicles is crucial to its goal to achieve a more sustainable energy future. Tesla makes significant progress in its goal to reach full autonomy, while improving driving experiences for all.

Grid-Powered Electricity

Tesla's Role in Revolutionizing Renewable Energy

Energy Storage

Tesla often brings to mind electric cars when people hear the name. Tesla is, of course, much more than an automobile company. Tesla is a leading company in the fields of renewable energy and storage. It is also committed to helping people transform the ways they generate, store, or use energy.

Tesla Solar Roofs, a New Era in Renewable Energy Generation

Tesla Solar Roofs are a revolutionary technology that has the potential to change how we view renewable energy. The Solar Roof is a revolutionary product, combining a beautiful and sleek design with advanced solar technology. This solution not only benefits the environment, it also helps your wallet. Tesla hopes to provide solar energy to anyone, regardless of architectural style or geographic location.

Solar Roofs is unlike any solar products available today. The tiles are made of durable tempered glass and engineered so that they look exactly like the traditional roof materials. The tiles have been designed to be extremely durable, to endure even the worst weather conditions. It's the integrated photovoltaic cell that makes the solar roof so magical. Solar Roofs' cells convert sunlight into electricity, which can power your house.

Solar Roofs offer the benefit of total invisibility from the ground. Solar Roof was designed to look like your current roof, unlike traditional solar paneling that may be unattractive and take away from your home's appearance. At a glance, the Solar Roof appears to be a standard roof. But it is much more.

Solar Roofs capture more sunlight than solar panels. Because the solar tiles are designed for maximum production of energy, they can harness sun energy when the sun does not shine directly on top. Solar Roof will continue to work hard even when the sun is not directly overhead.

Solar Roof's connection to Tesla's Powerwall storage system allows you to save the energy generated and utilize it at a later date. The Solar Roof gives you greater control over energy consumption. This will reduce your carbon emissions and help save you money on energy bills.

The Solar Roof has a significant impact on the environment. Tesla makes renewable energy cheaper and easier to access, which helps reduce our dependence on oil and creates a more green and cleaner environment. Tesla's Solar Roof represents just one way it leads the shift to a sustainable, energy-efficient future.

Tesla's Solar Roof represents a major breakthrough in the world of renewable energies. Solar Roof, with its efficiency and integration into Tesla's Energy Storage system, is revolutionizing solar energy. Solar Roof is an excellent solution if your goal is to reduce carbon emissions, lower your energy bill, or just have a stylish and functional roof.

Energy Storage Solutions to Make Clean Energy More Affordable

You may have wondered about how you can use renewable energy even when the weather is bad. Tesla's solutions for energy storage come in handy. Tesla's dedication to renewable energies extends far beyond its electric vehicle fleet to include the entire energy grid. The company's focus has been on developing innovative energy storage and management solutions that make renewable energy available, reliable and accessible 24/7.

Tesla's storage solutions include Powerwall, Powerpack, Megapack. This allows individuals, businesses and utilities to use excess solar or wind energy when needed. Powerwalls, such as the one designed by Powerwall, are rechargeable lithium-ion batteries for residential usage. This battery can store as much energy as 14 kilowatts-hours, and provide power backup during outages. Powerpacks or Megapacks on the otherhand are intended for large-scale commercial projects and utilities, with capacity ranging between 100 kWh and hundreds of MWh.

Tesla's solutions for energy storage not only assist households and businesses in reducing their dependency on traditional energy sources but they also aid the grid integration of renewable energies. In storing energy that is not needed, utilities can smooth out variations in the renewable energy's supply, making it reliable and more accessible. Not only does this reduce greenhouse gases, it also helps reduce the need for transmission lines and new power plants. This makes the energy grid efficient.

Tesla's storage solutions are already in use in more than forty countries. Clean energy is now available for all. From providing power to homes in Australia to supplying energy to remote areas in the US. Tesla installed the largest lithium battery on the planet in South Australia. It has the ability to power over 30,000 houses during blackouts. The large-scale battery project was not only a demonstration of the viability but also of the possibility to integrate renewable energy in the grid.

Tesla's innovative energy storage products are essential to its goal of accelerating the transition to sustainable energies. In providing efficient and reliable storage of clean energy, the products make renewable energies accessible for everyone. This reduces our dependency on fossil energy and supports the growth and development of a cleaner more sustainable grid. Tesla's continued investment and innovative efforts in this sector will allow it to become the market leader for energy storage. This is a great step forward in the development of clean energy.

Tesla's contribution to renewable energy: leading the charge

Tesla Solar Roofs as well as its energy storage products are having an impact on renewable energy. Tesla encourages the use of clean renewable energies like wind and sun by offering innovative solutions to generate and store energy. Tesla leads the charge to increase access to clean energy for all. Tesla provides products that are quick and easy to install. This makes it possible for anyone to produce and store energy.

Tesla's impact on the renewable energy industry and energy storage is not to be underestimated. Innovating and being sustainable is what the company stands for. They are changing the energy industry and the way people use and store energy. Tesla is making a huge impact in the world of energy. Whether you're an electric vehicle enthusiast, someone interested in renewable energies, or just someone who cares for the planet, its influence should be celebrated.

Tesla's Master Plan

Tesla's Master Plan Part 1 (A Vision for Sustainable Future)

Tesla's Master Plan 1 describes the goals and future ambitions of the company. Elon musk, Tesla CEO, wrote this document in 2006, and the company continues to work hard since that time in order to meet the objectives set forth in it. The Master Plan 1 of Tesla, the implementation and impact that it had on automotive technology are discussed in this section.

Tesla's plan for the first half of his masterplan had four primary goals.

Develop a low volume sports car which could demonstrate the viability of electric vehicles.

The money generated by the sport car could be used to build a cheaper car.

The profits generated from this affordable car can be used for the development of an even cheaper car, and technology that will make electric cars mainstream.

-Achieve clean and sustainable power through the use of solar products.

Tesla was required to establish a production company capable of manufacturing high-quality vehicles on a mass scale. The company had to make significant investments into research and technology, and also create a reliable supply chain of batteries and other parts. Tesla Roadster's predecessor, Model S, and Model X, were introduced by the company in 2008. Model S is one of most luxurious EVs in the world and has received many accolades.

Tesla's Model S was introduced in 2012. This full-sized luxury sedan could travel 300 miles per charge. Model S is a popular vehicle with customers and has been praised by many for its technology, performance and impressive design. Tesla continued to develop more affordable models thanks to the success of its Model S. It introduced the Model 3 in 2016 which targeted the mass market. Model 3 has quickly become one of most popular EVs worldwide and is helping to speed up the adoption rate of electric vehicles.

Tesla also works to offer sustainable energy to its consumers. Powerwall's Powerpack system allows customers to utilize clean energy without the need for sunlight. Tesla created the Solar Roof system that integrates solar cells into the roof. It allows for clean energy to be generated while improving aesthetics.

Tesla achieved a lot with its first plan. They have made great progress in developing their electric vehicles, as well as sustainable energy solutions. Tesla has created a standard of excellence in terms of quality, design and performance.

Tesla's Master Plan 1 document is a groundbreaking visionary document which has put Tesla in a position to revolutionize automotive technology. The Master Plan 1 has played a key role in helping Tesla reach its objectives and had an impact on industry. Tesla is a leading company in its industry because it has devoted itself to sustainability, affordability and innovation.

Tesla's future vision is constantly evolving, with the master plan of the company still being a very active work in progress. Tesla constantly works to improve their products and technologies.

Tesla Master Plan, Part 2: Expansion & Technology Improvement

Elon Musk's Tesla Master Plan Part 2 is a second chapter of Musk's vision. Published on Tesla blog July 20, 2016 it expands the original Master Plan 2006 Elon's Master Plan outlines Tesla's goals and objectives for the next decade. Four main principles are at the core of this plan:

Electric Vehicle Product Line Expansion

* Generation and storage of energy

* autopilot capabilities

Chapter 5: Tesla Is Expanding Its Reach

Expanding Tesla's Electric Vehicle Line- One of Master Plan Part 2's key goals is to extend Tesla's electric vehicle line. This will include the creation of new electric models, such as Model 3 Model Y or Cybertruck. Elon Musk said that the company's plans include a new electric sports car and semi-truck. Tesla's vision is to have electric vehicles available to the widest possible audience.

Tesla is focusing its efforts on energy production and storage, which is a key part of Master Plan Part 2. This company has a mission to help accelerate the shift to clean, sustainable energy. To do so, it offers products that store and produce energy. Tesla Solar Roof is an energy-generating roofing system. The Tesla Powerwall, on the other hand, is a battery for storing excess solar energy. Tesla Megapack - a massive energy storage system suitable for utilities and business - is another product developed by the company.

Tesla's Autopilot Capabilities is another important goal in the Master Plan Part 2: to enhance Tesla's ability to autopilot. Tesla's cars come with cameras and advanced sensors, which enable them to offer advanced features for driver assistance. Autopilot is being improved continuously to help the company achieve their goal, which is making autonomous cars a possibility. Tesla's Master Plan Part 2 details its vision of autonomous vehicles, which are able to be summoned by pressing a button. The plan is key in the future plans of the company.

The Master Plan Part 2 summarizes Tesla's plan to extend its reach into new markets. Tesla is making its electric cars and energy products as accessible as possible by expanding to other regions and new markets. Tesla is dedicated to making sustainable energy more available and accessible.

Tesla Master Plan Part 2 is an extensive vision for Tesla, Inc. The plan outlines Tesla Inc's objectives and priority, and is testament to its commitment towards advancing sustainable energy. The Master Plan Part 2 is a thrilling and ambitious vision that will shape the future of automotive technology, whether or not you are a Tesla enthusiast.

Sustainable Energy for All of Earth: Master Plan, Part 3.

Tesla CEO Elon Musk revealed their Master Plan Part 3 on Tesla Investment Day 2023.

Musk has presented Tesla's perspective on Earth's energy transition. From a fossil fuel-run world, to a Electrified Clean Energy Economy.

Then he said, "Over 80%" of Global Energy came from fossil fuels. Only a third provides useful energy or heat. The global energy systems is inefficient.

Globally, primary energy consumption today is approximately 165PWh/year.

The overall primary energy requirement can be decreased by converting all Primary Energy into sustainable sources, and improving the system efficiency.

Then, what are some ways that Master Plan 3 could be implemented. Musk has discussed the importance of large-scale energy storage solutions in order to facilitate the wide adoption and use renewable energy resources like wind or solar. These energy sources produce only as much power when the sun or wind shines. For a constant supply of energy, we must be able to save excess energy generated during periods of strong wind or sunlight to use at times when the energy production rate is lower.

Elon believes that, to reach this goal, we'll need huge amounts of batteries, around 24 terawatts hours. That is a large amount of power. These would include stationary Storage solutions like battery stations and large plant/stations.

Even though the number is large, this energy storage can still be achieved. And if the manufacturing facilities of batteries, electric cars and solar or wind farms are included, as well as mining for materials and the recycling process, they would all cost approximately 10 trillion dollars. Although it may seem expensive, this is a relatively low cost compared to that of the entire global economy. If you consider the 10 year period, the annual costs are less than 0.5% or 1% (based upon the GPD in 2022). Musk said that the investment expected in Fossil Fuels over the next two decades at the 2022 rate will amount to USD 14 Trillion. This represents a 40% increase compared to what can be done for a transition to a more environmentally friendly Energy Economy.

Further, an electrical economy would use about half the amount of energy as one based on combustion. The total energy needed to generate would therefore be lower. The production of wind and solar power requires relatively small amounts of land. They take up less that 0.2%. This source can produce significant energy just from the sun. It is surprising how much energy the sun can provide. You can get a continuous power of 25% by multiplying the aforementioned Gigawatts per Square Kilometer.

It is possible to move away from fossil fuels by converting to an electrical economy. This will allow us to achieve a more sustainable and energy-efficient future.

Energy efficiency is one of the main benefits that this transition brings. It reduces total energy consumption by a considerable amount.

Plans to end fossil fuel use

The goal is to displace fossil energy and replace it with renewable sources of power, according to what was previously discussed. Tesla, Elon Musk and their team have broken this plan down into five main pillars.

Renewable Energy can be used to power the grid.

The total energy consumption is 46 Peta Watts per annum, or 35%.

Solar and wind energy will help power the new electrified grid. The US Grid will add 60% more solar-powered generation in 2022.

By 2022 the solar energy deployment has grown by 50 percent per annum. This shows that the availability and affordability for renewable energy technologies are driving a growing trend towards more sustainable energy.

Additionally, wind turbines may be installed both onshore and offshore. They can be located in areas where there is less wind resistance. As the Earth is largely covered by water, we are able to reach full sustainability on a very small area of land.

It is anticipated that the investment required to switch from fossil fuels to renewable sources will be approximately USD 0.8 Trillion.

Switch to Electric Vehicles

21% of total: 28, Peta Watts per annum

It is clear that the trend toward electrification will not abate. Electric vehicles are one such example. They could alone lead to 21% less fossil fuels being used. Tesla, as well as other companies involved with this initiative are heavily invested in it. In 2022, EV production growth will be 59%.

EVS currently hold an impressive market share of 10 percent, marking a milestone. It is no secret that the transition from gasoline to electrical and automated vehicles has been rapid.

You should note that the assumptions made regarding the battery requirements for this switch are somewhat conservative. With the increasing autonomy of the fleet, there will be fewer vehicles and a lower battery requirement. Electric vehicles have reduced fossil fuel usage by 21 percent.

Tesla, as well as many other companies are actively involved in these activities. The result is a growth of 59% per year for the total production of electric cars in 2022. Electric vehicles have now taken up an impressive 10 percent of the global market. This milestone is exciting and significant. The old gas cars that are not autonomous will eventually be replaced by newer fully-electric and self-driving cars. This transition will be significant, just like the change from horses and cars to smartphones. This assumption is conservative because the autonomous fleet's benefits will reduce the number of vehicles that require batteries.

A goal of 20 million vehicles will be produced annually. As autonomous cars become more common, the need for personal vehicles will decline, although the precise number is still being debated. Globally around 2 billion automobiles and trucks operate. An autonomous Fleet will require fewer vehicles. In the presentation, a fleet consisting of approximately 1.4 billion automobiles is shown. This number is much smaller than it currently stands.

About 85,000,000 vehicles are made annually to help facilitate the transition to a future of autonomous and electric cars.

It is easy to acquire raw materials. For the electrified economies, we will end up using less energy to extract minerals. Sustainable energy would use materials that are needed to produce sustainable energy instead of fossil fuels, reducing the minerals mined. This is achievable, and all the resources needed to produce sustainable energy are readily available.

Although there is a current need to mine resources, it does not cost astronomical amounts of money for the materials needed for the transition into a sustainable energy-based economy. In the past, it has been shown that resources are found more as they are extracted. Also, since 2000 there is a greater availability of these materials.

Lithium is a common element on Earth, but there seems to still be some confusion. The element lithium is found in abundance on Earth. There are no countries that have a monopoly. It is estimated that the United States contains enough lithium for the whole planet. Limiting factors are the cost of refining and converting lithium to lithium carbonate or battery grade hydroxide. This is also true of other materials necessary for sustainable energy, and refinement costs are low. The need for them to occur is not difficult.

Nickel, however, is not the only problem. The world also has plenty of nickel and its reserves have increased. The iron cells will handle the majority of heavy-lifting for the electrification. And since iron is the commonest element on Earth we'll have plenty.

The electric vehicle, for instance, uses less fuel and raw materials than an internal combustion engine. The production of solar panels, wind turbines and other renewable energy sources requires less material than that used in traditional power plants.

Tesla Model 3s are four times more fuel-efficient than Toyota Corollas from the well all the way to the steering wheel. Electric vehicles are highly efficient. A Tesla Model 3 is four times more efficient from the well to the wheel than a Toyota Corolla.

The model 3 has the ability to go over one mile by using the amount of power that is required to cook pasta. Model 3 is a 4000 pound vehicle that requires very little energy.

Around USD 7 Trillion is the total estimated investment to switch from gasoline to electrically powered vehicles.

Chapter 6: Choose Heat Pumps

The total energy consumption is 29 Petawatts/year, or 22%.

It is important to decarbonize our economy by electrifying transportation buildings and industrial processes. It's good to see heat pumps being adopted more widely. They are important technologies for reducing the use of fossil fuels within buildings. Musk has mentioned that Tesla is also interested in getting into the heat pump business, to help with the transition.

Manufacturing Investments are estimated at USD 0.3 trillion.

Hydrogen High Temperature Heat Delivery

(17% of total: 22.2 Peta Watts per year)

Electricity in industrial processes at high temperatures is more complex, as you've mentioned. But there are solutions that can be found using specially-built equipment.

In addition to the green hydrogen you mention, it may play a part in decarbonizing certain industrial processes. Decarbonizing the economy will require various solutions.

It is estimated that the investment total on these items will be approximately USD 0.8 Trillion.

Sustainable fuel planes and boats

Total: 7 Petawatts/year - 5%

It will become a normal process as the EV Industry grows to apply the Engineering Learning Processes to

Planes and Boasts Manufacturing Design and Processes.

It is anticipated that the manufacturing investments on planes and vessels with EV technologies will total around US$ 1 trillion.

Tesla's Investor Day attracted investors, enthusiasts and the media. The event showcased Tesla's plans for the near future. Musk's focus on sustainability and corporate social responsibility, was another highlight at the event. This showed that Tesla, more than a company of automobiles, is a pioneer in creating a better and cleaner future.

Expanding ReachTesla's Global Growth and Impact on International Markets

Tesla's mission to revolutionize the world's transportation and energy systems has been embraced by people all over the world. From the United States to Europe to Asia, Tesla's commitment to sustainable energy and innovative technology is making waves and changing the status quo. The story of Tesla's global growth is a testament to the company's determination, innovation, and impact on the world.

Breaking into International Markets

When Tesla first started selling its electric vehicles outside of the United States, the company faced a number of challenges. Different cultural norms, regulations, and languages presented obstacles that the company needed to overcome. However, Tesla's relentless pursuit of its mission and its commitment to sustainable energy made it clear that these challenges were not insurmountable. The company's determination to succeed in international markets has paid off, and Tesla is now a major player in the global automotive and energy industries.

Tesla's Impact on the European Market

Europe is one of Tesla's largest international markets, and the company is having a tremendous impact on the region. Tesla's electric vehicles are changing the way that people in Europe think about transportation, and the company's Solar Roofs and energy storage solutions are making clean energy more accessible and affordable. Furthermore, Tesla's presence in Europe is increasing the demand for electric vehicles and clean energy solutions, which is having a positive impact on the European economy.

Expanding into Asia

Tesla's expansion into Asia has been one of the company's most exciting developments in recent years. With Asia being the world's largest market for electric vehicles, Tesla is well-positioned to take advantage of this growing market. The company has opened factories in China and is actively working to establish a strong presence in other Asian markets such as Japan and South Korea. By expanding into Asia, Tesla is helping to increase the demand for electric vehicles and clean energy solutions in the region, which is playing a major role in the transition to sustainable energy in Asia.

Global Reach, Global Impact

Tesla's global reach is having a profound impact on the world. The company's electric vehicles and clean energy solutions are changing the way that people think about transportation and energy, and it is helping to accelerate the transition to sustainable energy. Whether you are in Europe, Asia, or anywhere else in the world, Tesla's impact on the automotive and energy industries is something that should be celebrated.

Tesla's global growth is a story of determination, innovation, and impact. From Europe to Asia and beyond, Tesla is changing the world and revolutionizing the way that people think about transportation and energy. If you are a fan of electric vehicles or clean energy, you should be proud of Tesla's global growth and the impact that it is having on the world. The company's commitment to sustainable energy and innovative technology is making a difference, and it is something that we can all be excited about.

Charging Forward

The Development and Growth of Tesla's Supercharger Network

Tesla Motors has always been committed to making electric vehicles mainstream and accessible to all. However, one of the biggest challenges faced by early electric vehicle owners was the lack of convenient and reliable charging options. This is where the Tesla Supercharger network came into play.

The birth of the Supercharger network in 2012 marked a significant milestone in the evolution of electric vehicles. Tesla wanted to provide its customers with a fast and easy way to charge their vehicles while on long road trips, so they could enjoy the full benefits of electric vehicle ownership without having to worry about range anxiety.

The first Superchargers were built along popular travel routes in California and were designed to provide up to half a charge in just 30 minutes. This was a revolutionary step forward compared to the hours-long charging times offered by public charging stations. Tesla's focus on the customer experience was evident in the design of its Superchargers, which were not only fast, but also simple and stress-free to use.

As Tesla's vehicle lineup expanded and the demand for Superchargers grew, the company continued to invest in the network. Today, the Tesla Supercharger network is one of the largest and most extensive electric vehicle charging networks in the world, with over 20,000 Superchargers in over 2,000 locations globally. This makes it possible for Tesla owners to travel coast to coast in the United States and across entire countries in Europe and Asia with confidence.

But Tesla's dedication to the Supercharger network doesn't stop there. The company continues to innovate and improve the charging experience for its customers. For example, the recently introduced V3 Superchargers are even faster than previous versions, providing up to 200 miles of range in just 15 minutes. This is just one example of Tesla's commitment to delivering the best possible driving experience to its customers.

Tesla Supercharger network has been a key factor in making electric vehicle ownership more convenient and accessible. It has helped to eliminate range anxiety and has made long-distance road trips a reality for electric vehicle owners. As Tesla continues to lead the charge in the transformation of the automotive industry, it's exciting to think about what the future holds for the Supercharger network and how it will continue to make electric vehicle ownership even more accessible and enjoyable.

The Ins and Outs of Tesla's Incredible Supercharger Network

Have you ever wondered how Tesla drivers are able to travel long distances with ease? The answer lies in the heart of the Tesla Supercharger network. This innovative charging solution is a key factor in making electric vehicles a viable option for everyday use. In this chapter, we will dive into the technical details of the Supercharger network and explain how it works in a friendly and easy-to-understand manner.

What is a Supercharger, Exactly?

Think of a Supercharger as a high-speed pit stop for your Tesla vehicle. It is a charging station designed specifically for Tesla cars, providing a fast and convenient way to refuel your vehicle's battery. Unlike traditional charging stations that can take several hours to charge a vehicle, the Supercharger network provides a quick and efficient solution.

Each Supercharger station is equipped with multiple charging ports, called Superchargers, which can deliver up to 250 kilowatts of power to your vehicle. This means that you can charge from zero to 80% in just 30 minutes, making it possible to travel long distances without having to spend hours waiting for your car to charge. How cool is that?!

How Does the Supercharger Network Work Its Magic?

The magic of the Supercharger network lies in its use of high-power charging technology. Unlike traditional charging methods that use alternating current

(AC), the Supercharger network utilizes direct current (DC) fast charging, which is much more efficient.

When you pull into a Supercharger station and connect your Tesla vehicle, the charging process begins. Your car's onboard charging hardware, called the charge port, converts the high-power DC electricity into low-power AC electricity, which is then used to charge the battery. The charging process is monitored by the vehicle's charging management system, ensuring that your battery is charged safely and efficiently.

One of the most exciting things about the Supercharger network is that it is connected to Tesla's energy storage system. This allows Tesla to store excess energy generated from renewable sources and use it to power the Superchargers. This not only reduces the Supercharger network's carbon footprint but also helps to make it a more environmentally friendly solution.

A Network That Keeps on Growing

Tesla's Supercharger network is constantly expanding, with new Supercharger stations being added all over the world. The company's goal is to provide Tesla owners with access to fast and convenient charging options no matter where they are, making it possible to travel long distances without worrying about running out of power.

The Supercharger network currently spans across North America, Europe, and Asia, and the company plans to continue expanding the network to cover even more regions in the future. And the best part? Model S, Model X, and Model 3 owners get free access to the Supercharger network, making it easy and affordable to charge their vehicles on the go.

In conclusion, the Tesla Supercharger network is a game-changer for electric vehicle owners. With its fast and efficient charging technology, constantly expanding network, and free access for Tesla owners, the Supercharger network provides a convenient and sustainable solution for electric vehicle charging. So the next time you're on a road trip in your Tesla, be sure to take advantage of this incredible network!

Chapter 7: The Future Of Mobility

Tesla's Vision for Sustainable Transportation and Its Impact on Society

When Elon Musk founded Tesla Motors, he had a dream of a world where transportation was no longer dependent on fossil fuels and where electric vehicles were accessible to all. Over the years, Tesla has made great strides in bringing this dream to reality, but the company is not stopping there. Tesla's ultimate goal is to transform the way we think about and use transportation completely.

One of the most significant parts of Tesla's vision for the future of mobility is autonomous driving technology. Tesla is a trailblazer in developing advanced self-driving systems, which have the potential to greatly reduce the number of traffic accidents, improve road safety, and give people more free time to spend on the things they love. Tesla's mission to bring fully autonomous vehicles to market is in full swing, with millions of miles of data being collected and analyzed every day to improve the system.

In addition to autonomous driving, Tesla is also passionate about making transportation more sustainable. The company is working hard to bring electric semi-trucks to market, which have the potential to significantly decrease emissions and make long-haul shipping more efficient. Tesla's electric semi-trucks will provide a cleaner, greener alternative to diesel-powered trucks, and they'll help lay the foundation for a future where transportation is powered by renewable energy.

Tesla's vision for sustainable transportation is not just good for the environment, it's also good for society. By reducing emissions and improving road safety, Tesla is helping to create a better, healthier world for future generations. The increasing number of electric vehicles on the road and the growing availability of charging infrastructure are just a few examples of Tesla's impact. The company's continued efforts to improve and innovate will only further its impact on society and the world.

At Tesla, they believe that the future of mobility is electric, sustainable, and self-driving. The company's unwavering commitment to this vision is what sets it apart and makes it a leader in the automotive industry. Tesla's journey to revolutionize transportation has been exciting to watch and the future looks even more promising. With Tesla at the helm, the future of mobility looks bright and sustainable, and the company is poised to continue shaping the world in a positive way for years to come.

Tesla is Adding New Models

Tesla's evolution is no different. Tesla made waves in the last few years with their innovative electric vehicles, and dedication to sustainable transportation. However, Tesla will not be slowing its pace anytime soon. It has several exciting new model in the works.

Tesla Cybertruck

Tesla is set to introduce the Cybertruck in the near future, an electric pickup that will be unique to the marketplace.

Tesla Cybertruck has been a hit in the auto industry. This futuristic pickup is unique and attractive. Tesla CEO Elon Musk first presented it to the public in November of 2019. It has since been the focus of a lot discussion and interest among Tesla fans.

Technically, Cybertruck represents a major achievement in automobile design and engineering. This pickup truck is unlike any other because it has an angular body. The design is not solely aesthetic. Cybertruck's stainless steel body and glass armor are designed to make the vehicle as durable and robust as possible.

Cybertruck features a dual-motor, all-wheel drive, system which offers incredible performance. Tesla states that their top of the line version will go from 0-60mph in only 2.9seconds, making it faster than sports cars. Cybertruck's towing capabilities are also impressive, reaching a max of more than 14,000 lbs.

Cybertruck's range will be up to 500 kilometres on one charge. That is plenty for most users. In addition, the truck's 100 kWh batteries will make it possible to quickly recharge.

Cybertruck's advanced features will include a 17 inch touchscreen display, a camera 360 degree system and a sophisticated autopilot. All of these features make the Cybertruck an excellent vehicle to use for many different types of driving - from long-distance trips to work.

Cybertrucks, the world's first autonomous trucks, are due to arrive in late 2023. But mass production begins in 2024.

Overall, I think the Tesla Cybertruck has a lot of potential to change the face of the automobile industry. Cybertruck, whether you're looking to buy a new vehicle or are just a truck lover, is worth watching.

Tesla semi truck

Tesla Semi is another new model in development. It's an all-electric trailer-trailer. Semi, with an advanced autopilot and low operation costs, is expected to revolutionize logistical industry. Tesla Semi should also be able to make a positive contribution to the environment. It will reduce emissions, and promote sustainability in the transport sector.

Tesla Semi Truck is an all electric, heavy duty truck of Class 8 that was launched by Tesla on November 17, 2017. Tesla Semi was designed with long-haul transportation in mind. It is highly efficient, secure, and environmentally friendly.

Tesla Semi features an electric powertrain. The Semi's four motors are completely independent, unlike diesel-powered trucks. They provide performance and efficiency that is unsurpassed. They provide instant torque and rapid acceleration, allowing for the truck's ability to hit highway speeds without a tractor in five seconds. Semi-trucks can reach 65 mph on an incline of up to 5%, which is steeper than the average diesel truck.

Tesla Semi has a great range. Semis can cover up to 500 kilometres on one charge with their full batteries, so they are ideal for longer trips. This truck battery can be charged to 80% of its capacity within just 30 minutes. It allows drivers to quickly charge during breaks so they can get back on their way in no-time.

Tesla Semi features advanced features of safety that are designed to keep both driver and road users safe. This truck has an Autopilot, which is a system of advanced technology that assists the driver with traffic navigation and reduces the chance of accident. This truck also has multiple radar sensors, which provide 360-degree views and assist in detecting potential roadblocks.

Tesla Semi, in addition to being designed to provide safety and high performance, is also highly efficient. Its low-center-of gravity design and aerodynamic shape maximize the energy efficiency of the truck, which reduces charging frequency and operating costs.

Tesla Semi trucks will begin mass production in 2024.

Tesla Semi Truck is innovative, exciting and a great addition to the heavy-duty industry. Semi, with its advanced safety and energy-efficient features, has the power to revolutionize and transform the trucking industry.

A new, compact SUV is also being developed by the company. This would be a cheaper and more practical alternative for drivers to use on a daily basis.

Tesla is on a promising path with exciting new models. Tesla has a commitment to sustainable and innovation transportation. The new cars will push the limits in what is possible for the automobile industry. If you own a Tesla or are just interested in what the future holds for mobility, then these models are definitely worth your attention.

How to overcome obstacles

Tesla's challenges and efforts to overcome them

There are obstacles in every company's path, regardless of its size and industry. Tesla is not an exception. The electric vehicle manufacturer has experienced its fair share difficulties since 2003. Tesla is one such company that has overcome many obstacles to become one of today's most innovative, influential, and successful companies.

Tesla is facing a major challenge in convincing people of its vision. In the early days of Tesla, electric vehicles were a concept that was relatively new, with many people being skeptical as to their usefulness and reliability. Tesla needed to do a lot of work to dispel these misconceptions, and to show that electric vehicles weren't just possible but wanted.

Tesla is also facing production and logistics issues. In order to maintain its growth, Tesla has had to increase production to match demand. As a result, there have been delays and supply bottlenecks. In order to keep up with the growing demand for the vehicles it produces, the company had difficulty scaling its manufacturing capacity. As a result, production has been slowed down and there were temporary closures.

Tesla is a company that has overcome many obstacles despite its challenges. Its innovative spirit and drive to succeed have allowed it to continue to move forward. The company made major investments in their production processes to increase efficiency and streamline the supply chains. These have helped them overcome many of its previous production and logistics issues.

Tesla was also proactive about addressing critics' concerns, and invested heavily into research and development in order to enhance the reliability and performance of their vehicles. Tesla took several measures to improve customer service, such as the Supercharger network that makes charging Tesla vehicles easier.

Tesla has demonstrated a dedication to sustainability by reducing its environmental impact. This has enabled it to overcome many challenges. The company is investing in renewable energy technologies and energy storage. They are also committed to reduce their carbon footprint, as well as creating a better, sustainable future.

Tesla was faced with many challenges when it first started, but the company has persevered to become one of today's most innovative and influential businesses. Tesla has a commitment to sustainability, innovation, and will keep moving forward with its future.

Tesla Inc. Legacy and Impact on the Automotive Industry

Elon Musk was hired by Tesla Motors as a CEO in 2004. No one could have foreseen the dramatic impact of the company on the global automotive industry. Tesla has become one of world's most innovative companies.

Tesla began with one goal in mind. It was to help accelerate the shift to renewable energy. The company launched a range of electric vehicle that has redefined performance and style, as well as the possible limits of an electric car. Tesla Roadster, Model S, Model X, Model 3, and Model Y all set standards in electric vehicles. They also paved the path for a whole new generation of environmentally friendly transportation.

Tesla, in addition to its vehicles, has invested heavily into renewable energy sources and energy storage. These investments have further cemented the company's dedication to creating a more sustainable future. Tesla, with its self-driving car technology and growing Supercharger network is committed to making electric and sustainable transportation available for all.

Yes, there have been challenges along the way to revolutionizing automobiles. Tesla has had to overcome a lot of obstacles on its way. But, Tesla has always remained committed to its mission, its values, its passion, and pushed forward.

It is evident that Tesla Inc. has been able to achieve much more in its history than building electric automobiles. Tesla Inc. has revolutionized the automotive industry with its constant pursuit of innovation. Its commitment to sustainability and its unwavering focus on it will be felt for years to come. For years to come the company's impact will be felt on society, environment, and will leave a legacy of bravery, determination, and relentless drive for a more perfect world.

Chapter 8: Tesla Inc Was Founded By the Founder

Elon Musk is not the founder of Tesla. It was Elon who made the company his own, first investing in it and later directing its development from niche luxury to mass manufacture, as well as bringing in a solar energy company and promoting self-driving cars.

Tesla was in fact the fourth CEO of the company when the Tech Tycoon, now the wealthiest person on earth, took the job. This happened back in October 2008.

Martin Eberhard (founder of Tesla) and Marc

Tarpenning was interviewed by CNBC. She shared her memories on what it felt like to create and deliver the very first Tesla Roadster vehicles. What it took to convince everyone that electric cars were as exciting as sport cars. And how she brought Elon Musk onto the project.

Elon had first met them at a Mars Society member's luncheon. They were connected by their interest in the space industry. Musk was not yet SpaceX's founder when they met.

Eberhard had been "voted off of the island," according to him, following a lengthy legal dispute and subsequent settlement. Yet he still believes in Tesla's electric vehicles, considers them important for the safety of our planet, and is a Tesla stockholder.

Eberhard, a passionate electrical engineer, works to develop technologies which will allow electric car battery prices to be lower than what they are currently, but without any loss in safety, performance, or quality.

Tarpenning claims he has a continuing relationship with Musk and left Tesla at the time they were building Tesla's flagship Model S. "I have no regrets," he said, looking back. "From the very beginning, the whole thing was terrific. "It was both the worst and best." The plan has worked great.

Elon Musk has become a household name as CEO and founder of SpaceX, a private space venture. He is also the Chief Executive Officer (CEO), of Tesla Inc. Musk

PayPal was co-founded by him (PYPL). He also invested in a number of digital startups. And in April 20, 2022, discussions began for an agreement to privatize Twitter Inc. It is not surprising that his success and flair are reminiscent of other colourful tycoons in U.S. historical history. With an estimated $220 Billion net worth as of June 20, 2022, he has become the richest person on the planet.

Musk acquired his status first in 2021 after surpassing Amazon.com Inc., (AMZN) founder Jeff Bezos.

Elon Musk was born and raised in South Africa, and spent time living and working in Canada before moving to the U.S. Musk, born and raised in South Africa before moving to America, spent some years in Canada.

Elon Reeves-Musk was born in Pretoria South Africa in 1971. He is the third child. His mother was both a Canadian and Canadian-born model, as well as a

South African engineering graduate. Musk spent most of the time with his father after his parents divorced in 1980.

Musk's escape was technology. In the early days of home computers, he learned programming on a Commodore VIC-20. Musk was already proficient in programming when he designed Blastar-- a game that resembled Space Invaders. For $500, he sold the BASIC codes for Space Invaders to a magazine.

Musk and Musk's younger brother were planning to construct an arcade video games near their childhood home.

school. The parents of the children voted against it.

Musk started his computer entrepreneurship career after

completing his physics degree at University of Pennsylvania. Early successes included Zip2 & X.com which merged together with a PayPal-like business.

Later he said, "(Physics') foundation is good for thinking." "Boiling things down to the truths at their heart and reasoning up from there" is what he said.

Musk, along with his younger sibling Kimbal, founded Zip2 in the year 1995. Zip2 is a startup web software that helps newspapers produce online city guide.

Compaq Computer Corporation purchased Zip2 from Compaq in 1999. The purchase price was $341 million. Musk spent the Zip2 purchase money on X.com – a fintech business before the term became popular. PayPal emerged when X.com merged with Confinity a service that transferred money.

Peter Thiel sacked Musk as PayPal's chief executive officer before eBay bought the payments startup (EBAY), but Musk still benefited because he owned 11.7 percent of PayPal.

Musk told an interviewer in 2018 that, "my PayPal profit after tax was about $180,000,000". In a 2018 interview, Musk said that $100 million went into SpaceX. $70 million was put into Tesla. And $10 million in SolarCity. And

You almost had to take out a loan for the rent."

Musk first became interested in electric vehicles as an early-stage investor back in 2004. He invested roughly $6.3 million and eventually joined a management team including Martin Eberhard. Eberhard left his position in 2007 after several conflicts. An interim CEO was then appointed, and Musk assumed the role of CEO and product architect. Tesla is now the most successful car company in history and one his most well-known companies.

Tesla has a significant presence in solar energy due to the purchase of SolarCity.

There are two solar rechargeable batteries available. Powerwall's smaller model is designed to provide backup power for homes and for use in off-grid situations, while Powerpack was created for larger businesses or grid-connected electric utilities.

Musk's early interest in science-fiction, fantasy, and philosophy is reflected in the idealism he has for human progress and his successful commercial career. The topics that he works on are those deemed essential for the human future. This includes the use of renewable energy sources and space exploration. Musk has resisted detractors, disrupted industries, and earned the most money anybody ever has from PayPal, Tesla Motors, SolarCity, and SpaceX--game-changers all, despite the inevitable blunders.

Elon Musk was the wealthiest person on Earth in 2022.

Chapter 9: Tesla Inc

Tesla was established in 2003. A group of engineers wanted to demonstrate that there were no sacrifices needed for consumers who chose to drive electric. Instead, they could enjoy a more enjoyable driving experience with electric vehicles.

Tesla today manufactures all-electric vehicles, as well as sustainable energy storage and generation technologies. Tesla's belief is that the faster the world transitions away from fossil energy and toward a zero-emissions future, the more it will benefit.

Tesla Inc., or Tesla for short, is a multinational automobile and energy corporation. It develops, designs, produces, markets, and leases energy generation and storage solutions, electric cars. Model Y, Model 3 and Model X are all sold by this company.

Cybertrucks such as Tesla Semis and Tesla Roadsters.

Tesla's automobiles are built in Fremont (California) and Gigafactory, Shanghai. Tesla has adopted a pro-active approach to safety in order to reach its aim of being the safest manufacturing facility in the entire world. This includes requiring all production personnel to undergo a multiday course prior to stepping onto the factory floor.

Tesla is now continuing to train employees on the job and constantly monitor their performance so they can make changes quickly. Tesla has seen its safety rates improve with increasing production.

Tesla offers end-toend energy solutions including generation, storage, consumption, maintenance and installation of energy systems. A network of independent shops and gallery-owned businesses advertises the company's vehicles and helps sell them. The company has facilities and activities in Asia Pacific and Europe, as well as the US, Germany and China. Tesla is located in Austin Texas.

As of 2020, this firm held the world's largest sales of electric battery cars. Tesla Energy's subsidiaries are a major installer and producer of solar energy systems in the United States. Tesla Energy also ranks among the leading global providers of battery-based energy storage, having deployed 3,99 gigawatt-hours in 2021.

Musk says that Tesla aims to hasten a shift towards sustainable transportation and energy through electric cars and the use of solar electricity. Tesla produced its first model of automobile, the Roadster, in 2009. Model S, Model X SUV and Model 3 sedans followed in 2012.

Model 3 became first plug-in car in history to sell one million cars worldwide. Tesla's global sales in 2021 were 936.222 automobiles, which is an 87-percent increase from the previous year. At the end of the 2021 fiscal year, cumulative sales had exceeded 2.3million cars. Tesla's stock market valuation surpassed $1 trillion on October 20, 21. They were the 6th company in U.S. History to reach this milestone.

Tesla offers four different automobile models as of 2022. Model S is the second generation Roadster. Model X is a semi-truck. Model Y, formerly known as first-generation Tesla Roadster has been retired. Tesla also has plans for the second-generation Roadster and semi-truck. It is also planning to build a pickup truck called Cybertruck.

Tesla Energy's subsidiary Tesla Energy sells and builds solar energy systems, battery storage devices, as well as related goods and services to industrial and commercial clients.

Tesla Energy has a number of products that generate electricity, including solar panels (made for Tesla by another firm), the Tesla Solar Roof System (a system using solar shingles), and the Tesla Solar Inverter. Powerwall (an energy-storage device for households) as well as the Powerpack/Megapack, large-scale storage devices are also available. Tesla Energy offers software designed to assist clients in monitoring and operating their installations.

Tesla's mission is to continue making affordable goods available to as many people as possible, ultimately accelerating the adoption of renewable energy sources and environmentally friendly transport. When combined with renewable energy and batteries as well as electric vehicles, these technologies will be even more effective. This is what we are striving for.

Tesla's Logo.

Automaker logos can and should be easily recognizable, even if they are not as distinctive as the cars themselves. This makes the design of a logo crucial. Tesla's icon is not anonymous. But it's actually more complex than you might think.

Tesla initially intended for the Tesla logo to sit within a symbol of a shield. However, they later decided to use the T as their trademark. Tesla CEO Elon Musk said the stylized "T", which appears as a logo, is in fact a representation of their goods.

Musk claimed that Tesla's logo represents the sectional cross-section of a motor.

Musk may be saying that the main part of "T" represents one of two poles that stick out from a motor rotor. A second line is added on the top to represent the stator.

Tesla has also created a logo that is opulent, strong and elegant. Its black and White color scheme makes this possible. Tesla will not be ready to start selling cars in the mass market until it has sold premium vehicles to consumers who could buy any other car. Tesla had to create a distinctive logo for this segment of the market, given its importance.

This communicates high-end luxury.

Tesla's elegant and futuristic logo expresses everything about their brand.

Tesla has certainly benefited from this highly successful logo.

Section 3.

Tesla Features

Tesla's cars come with many interesting features and easter-eggs. Elon Musk has a unique sense for humor, and these elements have led to Tesla's following. Teslas stand out for a variety of reasons, including their sentry and puppy modes as well as "caraoke" features.

1. Autopilot.

Autopilot comes as standard equipment on every Tesla car. It allows the car to drive, accelerate, or stop itself within its own lane. The technology does not take the place of a human driver. However, it is designed to simplify driving and decrease accidents.

Tesla says that its new models feature eight cameras with 12 sensors to provide 360-degree views of their surroundings. Tesla CEO Elon Musk has often equated cameras to human vision.

There is also a Full Self Driving $12,000 Add-On that allows the system to change lanes and recognizes stop lights or stop signs. Tesla customers who meet company requirements regarding driver safety can download the functionality, currently only in beta testing.

2. Caraoke.

Tesla is introducing a new function called "caraoke", which will allow Tesla owners to choose their favorite songs from the vast collection of lyrics and tunes.

China was the most popular country for this feature. In China in January, the in-car microphones for $188 sold out after an hour. TeslaMics is currently available only in China.

Caraoke music is available in many different languages. Do not think you can sing along to the songs while driving. The vehicle must be parked for the function to work.

3. Bioweapon defence mode.

Tesla Model X & Model Y along with more recent Model S & Model 3 models are equipped with a HEPA filtration to prevent hazardous substances infiltrating the vehicle's interior. The system filters everything, from bacteria and pollen to pollutants.

Tesla has produced a short film explaining how it can help protect their drivers against biological warfare weapons.

4. Touchscreen.

Tesla includes a touch screen that is unique to each model. It offers a wide range of services such as video games, streaming, traffic information, etc.

The panels also house conventional services such as temperature control and navigation, yet they look completely different.

5. The video game streaming industry is growing.

Tesla continuously upgrades its infotainment with YouTube, Netflix and Hulu.

Also included are options such as Toybox, Theater, Arcade and Browser. Drivers can play games using the buttons on the steering wheels or USB controllers in the Arcade option.

Although the vast majority of features is only available while the Tesla vehicle is charging, some drivers claim that the feature has been a great way to keep them entertained when they are stuck in heavy traffic or waiting for their Teslas to charge.

6. The Web Browser

Tesla's premium connection lets you access multiple sites while parking your vehicle, including Hulu Spotify, and PLEX.

7. Air suspension.

Tesla's feature a clever air-suspension system with automatic adjustment based upon the GPS coordinates.

Tesla users can also modify the stiffness directly from the control panel. You may find this feature useful, especially if you are driving across country and experience changes in your route.

8. Advanced parking sensors

Tesla drivers can be alerted to any potential danger by using the powerful sensors that are built into their Tesla vehicles.

You will be alerted by the car's visual and acoustic cues if an object approaches too closely to you.

9. Superchargers.

Tesla has more than 35,000 supercharging outlets around the world.

Tesla states that it will charge your vehicle within 15 minutes and allow up to 200 mile of driving.

In the recent past, EV users have argued fast-charging points give Teslas an advantage because of their speed as well as general availability.

Tesla Superchargers represent around 58 percent of US fast-charging ports, according to Department of Energy.

10. The Key.

Other than the usual Tesla keys, there are many other options.

For those who don't wish to use their phone or card, you can purchase a unique Tesla-inspired keyfob.

11. Auto-presenting doors.

Tesla Model X's doors will unlock and open when the driver holds the key.

Tesla displays the automatic-opening feature on their in-car screens.

Musk referred to the Model X function as "one of the most amazing" features in July.

Chapter 10: Tesla Models

You can choose from four models currently available for sale.

Tesla Model S.

Prices start at $99,990/PS91,980.

Availability: US, UK.

Model S resembles a Jaguar executive saloon with similar low and long design lines. Model S helped Tesla gain recognition as a manufacturer in 2012. Tesla has continually improved the Model S ever since it's introduction. There are two current versions of this car: Model S Plaid. Recent years have seen the Long Range Standard Performance Model names being used.

Model S models older than the current model are rear-wheel driven, but the latest Model S's have AWD. Plaid, a brand-new model and the new flagship of the Model S range, has a 0-to-60-mph time under 2 minutes. It is powered by 'tri-motors' (rather than dual motors) which are superior to those found in the Model S.

Autopilot is included in every model. The driver can choose to upgrade the system to include navigation and other features. In the near future, autosteer will help you navigate city streets. Enhanced summon is also coming. Your parked car can find you in an underground parking garage.

Long Range: AWD. 375 miles. 0-60mph takes 3.1 secs.

Plaid: 348 miles AWD. 0-60mph takes 1.99s.

Tesla Model 3.

Starting Price: $46,990/PS45,990/EUR42,990 US, UK, Europe.

Model 3 launched in the US for the first time in 2016 and was positioned as an inexpensive alternative to Model S. Prices have since been reduced making the Model 3 the most affordable Tesla. Model 3 has become extremely popular, and you will see it on many roads.

Model 3 began as a four-door sedan. Today, it is available in just three options. All ranges are determined by adding up the numbers. While Autopilot is available, a complete upgrade to self-driving costs around $12,000/PS6,800.

Model 3: RWD, 267 miles, 0-60mph in 5.8 seconds.

Model 3 Long Range: AWD, 334 miles, 0-60mph in 4.2s.

Performance: AWD. Drive 315 miles and reach 0-60mph within 3.1 seconds.

Tesla Model X.

Prices start at $114.990/PS98.980

Available: US (including UK), Europe, and Australia.

Tesla Model X is "the soccer mom's solution" to electric cars. Model X, an SUV-styled EV with seating for seven, is similar to the Model S in terms of design, especially its interior display.

With its rear Falcon Wing door (like the Delorean vehicle in Back to the Future), and the big touchscreen, your kids will think you've bought an automobile from the future. The two models come in three versions: the 5 seater (included), the 6 seater ($6,500 additional), and finally, 7-seater ($3,500 extra). It comes with Autopilot, and you can upgrade to self-driving for $12,000/PS6,800.

Model X: AWD, 329 miles, 0-62mph in 3.8 seconds

Model X Plaid: AWD, 311 miles, 0-62mph in 2.5s.

Model X variants have also been numerous, as with the Tesla Model S. But now, they've all been consolidated to just two: the Model X and Plaid — the latter including the trimotor-system.

Tesla Model Y.

Starting Price: $64,990/ PS54.990/ EUR56.990

The following countries are available: US, UK, and Europe.

Model Y fits in the space between Model 3 & Model X as a small crossover. There will be five seats and plenty of room. It is primarily aimed at people looking to save money on the Model X.

Tesla is focusing on the market for small SUVs because it has the best sales at this time. Tesla claimed initially that there would be four variants, but now only provides two.

Model Y Long Range: AWD, 318 miles, 0-60mph in 4.8s.

Model Y Performance: AWD, 303 miles, 0-60mph in 3.5s.

Tesla only sells a modified variation to Tesla Workers.

Cybertruck.

Prices starting at: $39900/PSTBC

Availability: expected in late 2020.

Tesla has shifted its focus to pick-up truck design, and the Cybertruck is a result. Reservations are still available for other areas, too. This vehicle is more likely to become popular in the US.

Cybertruck will feature a unique angular body, with a stainless-steel shell and protective glass. The truck also has 100 CUFT of storage (2830 Liters), as well as a 7,500+ pound towing capacity. Tesla says that the US will be the only market for these features, while other countries may have their own specifications.

Original suggestions for the Tesla Cybertruck included three variants:

Single Motor RWD 250+ miles. 0-60 takes 6.5 seconds.

Dual Motor 4WD: 300+miles and 0-60 in less than 4.5 minutes.

AWD tri-motor 500+ miles and 0-60 is 2.9s.

Production delays are expected until the end of 2023 for Dual- and Tri-Motor Models.

Roadster.

Starting price: $200,000, bookings available.

Expected to be available in 2023.

Tesla Roadster began its entire development process in 2008 and is expected to be available in 2020.

Roadster has the ambition to be the fastest vehicle ever on the roads. The Roadster boasts some amazing stats including a time from 0-60mph in 1.9 second and top speeds exceeding 250mph. This vehicle will be able to seat up to four people and has a removable canopy.

Reserved models include the Roadster as well as the Founders Series. Prices are $50k different between both. These are the main specifications:

Roadster: AWD 620miles, 0-60mph 1.9 second.

Section 5.

Tesla Careers.

Is Tesla your ideal company? This section is just for you. This is not a quick process.

Tesla offers many job options. A lot of people want to work for Tesla because it is so well known. The following information will help you if applying to Tesla is something you would like to do but you don't really know where to begin. In this chapter, we will look at the qualities and skills required for success in a job application.

It is important to know what the company stands for, its history and achievements before beginning your application.

What are the chances of getting a Tesla Job?

Tesla's hiring process is extremely difficult. Your resume must be unique. For preparation, you can also review the previous Tesla test questions and interview questions. Tesla's goal is to employ employees who have the same mission as it does in terms of design and sustainability.

Is a university degree required for employment at Tesla Inc.?

Musk issued another message, this time from Tesla Owners located in Austin. Musk said, "High School Grads. You don't need a university degree to get a job with Tesla. If you've graduated high school and want to join Tesla, that is possible.

What is the minimum qualification to work at Tesla?

Tesla is accepting of GED and High School Diploma as well as related post-secondary training in vehicles (or relevant military service). It is mandatory that you are at least 18 years of age. If you have an 85 percent GPA or higher, then you can enter each training session of 16 weeks and successfully complete it.

Tesla Employee Benefits - Summary

Health Insurance Life Insurance. Vision Insurance Dental Insurance. Accidental Death Insurance Reduction. Long-Term Disability Insurance.

Tesla's dream is worth following. Tesla has a mechanical firm.

Do not resist the urge to accept pressure.

Bring your best abilities to the table.

Make yourself known as someone who can change the game.

Variety has its own power.

Display your strength

Being busy counts.

Tesla Interview Procedure is divided into 4 stages: Cycle of Assessment: In the first round, employees will share their experiences and achievements with you. This interview will assess your technical ability. Normally, employers will interview five to fifteen applicants at this level for each position.

Elon highlighted the importance of four factors.

Tesla's workers had more opportunities to learn, their employment was more reliable, and they were given better stock options. Glassdoor's rating site gave Tesla a score of 3.4 out 5, indicating that employees, in general, enjoyed their work.

Interview questions:

'Why Tesla?.

"What motivates you in working on this field?"

How to Improve Business Culture

'What are your car knowledges?

Tell me about the problems you faced and how they were resolved.

Tesla takes into account diverse thoughts and characters. The ability to solve a problem is great, but the creative, unique way you approach the challenge (means), that represents a tremendous level of integration.

Impact. Tesla offers a job to anyone who has the ability to enhance and truly feel their preparedness.

Tesla's employees or former workers have a number of positive things to say about getting a Tesla position. This organization offers many ways to become familiar with regular labor. There are fantastic stock-options, fairness, and the possibility to challenge intelligent coworkers.

Benefits.

Tesla shares the same reputation as many large corporations for providing significant returns and payouts. Tesla provides a range of excellent benefits for its employees. It offers comprehensive benefits, including retirement plans and 401(k).

Tesla provides both short and long term disability benefits, along with a discount of Tesla's shares. It also offers to pay back tuition expenses for those who choose to further their studies. Tesla will take care of you if you are hired.

Tesla offers several exciting career options. Several of the positions available are open to those with only high school degrees, but others will require an advanced degree as well as substantial work experience. Every organization wants the best talent in each case. You must take courses in machine learning before you can become a Tesla engineer. For example, if you're interested in joining the automotive design department of the company, you might choose to enroll in a university-based program to learn about automotive engineering. Then you need to make sure that you highlight all of your skills in your CV. After you succeed in this regard, the only thing left to do is test and hire the domain.

Chapter 11: The Study On Aspecification

TESLA SEMI's SPECS

The Tesla Semi will come with 500-mile range as per Musk.

In the vicinity of close to the Tesla Gigafactory in Nevada, close to Silver Springs, the long-awaited Tesla Semi was just sighted.

A semi-truck that was pulling a trailer observed parked at a roundabout and then softly swiftly and ease, accelerating. It's to be a bit unbelievable, especially when you compare it to normal diesel-powered vehicles.

It is important to remember this: the Tesla Semi has been advertised for a 20-second 60-mph 0-60 percent (96.5

kilometers/h) acceleration speed (when completely loaded).

It is not clear about whether the trailer in the video of a few weeks ago was filled or not. Meanwhile, a different video of Tesla Semi, which could be the same model, was taken on the highway when it was passing the diesel vehicle:

In lieu of accelerating, the Class 8 electric truck's most important performance element is its capacity for load as well as its distance (as well as its quick charging) (which is a plus in addition to the standard specifications).

If things don't go according to the information provided in 2017 We'll need to wait for a while for more information about performance and features, the carrying capacity as well as autonomy. The most important one, however it has a range of 500 miles. Below are a few more recently released Semi data.

The Tesla Semi won't arrive in December 2022 when its original delivery date was scheduled for the year 2019. Although there's been an extended delay of three years Pepsi hopes to get the initial Tesla Semi on the road on December 1. Because no other company has developed an electric vehicle, Tesla is (unsurprisingly) far ahead of its competitors.

It's intriguing that it's one of the most affordable Tesla automobiles to date. In addition, it's sturdy enough to be able to serve for transportation. As per reports this car comes with the following attributes:

It is able to transport the biggest possible semi-truck payload of 80,000lbs

There is enough juice to speed up to 0-60 mph and a complete payload in less than 20 minutes

Full-speed motorways speed up an 5% grade when loaded with full loads

An battery 8.5 times bigger than the one from the Model S

A distance of 500 miles including a payload of full

Capacity to recharge up at 70% within 30 minutes

It is faster than gasoline-powered trucks however, it does not have the drawbacks that require stopping and take a charge for hours at a period of.

The new Tesla Semi is able to perform exactly the same tasks as the regular semis, however it does not contribute to the environmental pollution, and costing trucking firms hundreds of thousands of dollars in fuel. The brand new Tesla Semi can, however it will only be capable of charging these speeds when connected to a specific Tesla Semi charger.

It won't matter over the long term, due to the huge reach of Tesla. The launching of the Tesla truck is hugely successful except for a problem that is similar to Samsung's Note 7 explosion battery tragedy. It is possible that drivers will see a rapid growth in charging stations that is reminiscent of a wildfire.

Simple and detailed information on its FEATURES

Performance

Three motors independent provide rapid torque and powerful acceleration at any speed. This means drivers are able to merge in a safe manner and remain in sync with other traffic. Speed up from 0-60 mph within 20 minutes, with all the gear fully loaded and still maintain highway-level speeds climbing steep hills.

Maximum Safety

Semi comes with the latest safety functions that are coupled with the most advanced brake and motor control systems to provide traction and security in any weather. The central position of the seat gives an improved view of the driver and a fully electric design helps reduce the chance of rollovers as well as cabin intrusion in the event incidental accidents.

Massive Range

Combination trucks are responsible for around 18 % of U.S. vehicle emissions-- Semi will assist in reducing this. At less than 2 kWh for each mile Semi is able to be able to travel for upto 500 miles with just one charge. You can get up to 70% of your range in just 30 minutes with the Tesla Semi Chargers.

COIN OF OWNERSHIP

The average cost of a semi-truck is $150,000. It can go up to $200,000 if you are buying a luxury version. The Tesla Semi is currently going at $180,000. This is on the higher end of the scale, however the cost savings make the initial investment for fleet owners far more valuable.

The need for oil changes is not necessary with electric semis. The only reason brake pads have to be replaced occurs around once every one million miles if maintenance is necessary. Only things that require replacement are the wipers, tires as well as washer fluid. The cost to charge for the Tesla Semi is significantly less than that of diesel fuel. It is 2.5 times less than an average semi. After just three years of operating, it goes as it as to claim that the Semi could save around $200,000 in expenses. These savings would be massive for a huge fleet.

DURABILITY

A semi-electric model will be able to function with less maintenance interruptions thanks to an updated electric system. Tesla says that even after 200,000 miles its regular models will continue to run with a battery that is 90% capacity. There is a chance that the semi will not even begin losing battery power or capacity until it's been around 300,000 miles in accordance the battery's size.

However, decline in battery performance is gradual. In the event of at least 500,000 miles the vehicles should be completely operational. After that, engines for traditional semis will need to be brand new. A majority of fleet owners dispose of their vehicles and buy new ones for their replacements.

To power these fully battery-powered semi trucks, there's an additional issue, namely "Megachargers." This is what I learned about the subject just one month ago. More importantly, Tesla started taking orders for the Tesla Semi in May, and then in June, it became clear that Tesla built the Megacharger located at an Frito Lay as well as a PepsiCo manufacturing facility located in Modesto, California, in anticipation of initial Semi deliveries to consumers. This year, in January Megachargers were also set into place in Tesla Gigafactory 1 (Giga Nevada).

I'm sure we'll learn more soon regarding the specifications as well as features and charging stations that will be available for Tesla Semi. Tesla Semi, with deliveries commencing in less than 2 months. Keep an eye on the news!

Chapter 12: Conceptual Review

Reviewing some key specifics that were revealed at last year's Tesla Semi unveiling, let's look at what's changed, what's not, and the things that remain in the dark.

The most likely thing that drivers will be able to feel and see on the design front is like this:

One of the first things you will are likely to notice upon entering Tesla Semi Tesla Semi is the full revamp of the cabin. The cab has fishbowl-like views across the front and sides of the cabin, it places the driver at the center of. The gauges and audio, as well as navigation and temperature controls that are usually found inside the dash are now available on two 15-inch touchscreens mounted on the sides which were taken out of models like the Model 3 where standard trucks include the dash.

Below are a few specifications of characteristics, specifications, and contrasts of the 2017 electric semi-truck.

than a diesel truck by 20%. $1.51 per mile for diesel trucks. $1.26/mile for an Tesla Semi. Tesla will keep costs of electricity at a rate of 7 cents per kWh. (Note: operating cost of a diesel-powered truck in the present is not less than $1.51/mile.The company also says that it will not increase the price of diesel trucks.

The Tesla Semi is more than twice the amount in an entire convoy than a diesel car. Additionally the Tesla Semi is more affordable than trains and can currently be in use and is 10 times more secure when compared to driving. The price per mile is reduced to $0.85.

(Note: According to my view, "can be deployed today" implies that Tesla is running the software to support semi-autonomous vehicles designed to allow convoys to travel. Convoys made by Tesla Semis might be deployed prior to November 2017, since it was evident that the vehicles were not yet in production.It is important to note that the trucks are not yet ready for manufacturing.

The drag coefficient of that of the Tesla Semi is 0.36. Diesel truck: 0.65 to 0.70; Bugatti Chiron: 0.38. The bottom of the truck is completely smooth. When driving, a computer device draws the truck closer to the cabin.

500-mile range with highway speeds and the GVW. (More than 80 percent of journeys are less than 250 miles)

At a gradient of 5 that the Tesla Semi is capable of sustaining 65 miles per hour. If you climb 5 the diesel semi-trucks can only reach speeds of 45 miles per hour.

0-60 within 5 seconds in comparison to the diesel semi's 15 minutes. To go from zero to sixty in just 20 seconds for an average weight of 80,000kg.

The range of 400 miles can be charged in just 30 minutes using "Megachargers."

drivetrain with a million mile guarantee.

Note: Having heard that Tesla's electric motors will last for around 100,000 miles, I don't think the claim.It's a fact.

The Tesla Semi is able to run just two of the four motors thanks to their redundancy, and continues to speed up faster than a diesel engine.

Note: The Tesla's website at present shows only an tri-motor (3-motor) choice. The future may include the option of a 4-motor is likely to be announced.[Note: Tesla's website currently only shows tri-motor (3-motor) options.

The thermonuclear glass is explosion proof. The glass will be returned in full reimbursement if the glass survives an explosion of nuclear nature.

[Remark: Hmm. The current context this assertion seems intriguing.[Remark: Hmm.

www.ingramcontent.com/pod-product-compliance
Lightning Source LLC
Chambersburg PA
CBHW051246050726
47594CB00001B/332